CIRCULAR ENERGY

Dados Internacionais de Catalogação na Publicação (CIP)
(Câmara Brasileira do Livro, SP, Brasil)

Coletta, Osvaldo Dalla
Circular energy : clean, renewable and sustainable energy / Osvaldo Dalla Coletta. -- Santo André, SP : Ed. do Autor, 2023.

ISBN 978-65-00-59601-4

1. Desenvolvimento sustentável - Brasil 2. Energia - Fontes alternativas - Aspectos econômicos 3. Energia elétrica - Aspectos ambientais 4. Energia sustentável 5. Recursos energéticos - Brasil 6. Recursos naturais renováveis I. Título.

23-141108 CDD-333.794

Índices para catálogo sistemático:

1. Brasil : Energias renováveis : Desenvolvimento sustentável : Economia 333.794

Inajara Pires de Souza - Bibliotecária - CRB PR-001652/O

Graphic design and Desktop publishing: Eliane Otani/Visão Editorial

Osvaldo Dalla Coletta

Economista CRE – SP n. 6.813

CIRCULAR ENERGY

Clean, renewable and sustainable energy

CONTENTS

CONCEPTS

The main concepts used in this book are as follows:

- It is impossible to create or destroy energy.
- Nature always uses energy in a circular way, reusing it by transforming it, in a natural process, from one form into another.
- Circular energy means reusing the amount of energy used, minus the amount of energy dissipated in the process of transforming energy from one form to another - for example, from mechanical to electrical form.
- Using energy in a circular way is important for the environment because it is clean, renewable and sustainable; therefore, it does not pollute the environment with toxic gases

and does not cause any other kind of environmental impact.

- Currently, humans use energy only once and discard it, and thus need to produce more energy through means that cause strong environmental impacts. For example, the use of fossil fuels that release carbon dioxide gas into the atmosphere, the construction of hydroelectric power plants that alter the natural conditions of rivers and flood large areas of land, and the construction of nuclear power plants by nuclear fission to produce electricity, which generate radioactive atomic waste that is polluting and extremely dangerous, as well as presenting a serious risk of nuclear accidents similar to the explosions of nuclear bombs.

The author of this book hopes that the content presented here will intensify the use of energy in a circular way and that it will motivate readers to discover new possibilities for reusing energy.

REUSE OF ENERGY USED BY HUMAN BEINGS

- **In the public environment:** install hydraulic turbines coupled with electric generators in the master pipes of the water tanks of both public water distribution networks and public buildings, in order to transform the hydraulic force of the water descent into electric energy. Install micro hydraulic turbines coupled with micro electric generators in the pipes of taps, showers, washing machines and toilet flush boxes, in order to transform the hydraulic force of the water output into electric energy. Install piezoelectric carpets on all the floors of public buildings, as well as on the sidewalks of the streets of cities with a large flow of people, to produce electrical energy to be injected into the public electric power distribution network.

Replacing the counterweight (dead weight) of elevators with electric generators. An electrical generator is always on to balance the weight of the empty elevator with the effort it requires to be moved. The additional amount of electric generators to replace the elevator counterweight is determined by the capacity (kilograms) of the elevator divided by seventy-five, so that one electric generator is activated for every seventy-five kilograms of elevator load weight to be carried. Thus, for every seventy-five-kilogram decrease in the weight of the load to be carried by the elevator, one electric power generator unit is deactivated. The effort to move the electric generator that is always activated is equivalent to the weight of the elevator without load and the effort to move each of the additional electric generators is equivalent to seventy-five kilograms.

- **In the private commercial environment:** install propeller-less wind electric generators on top of commercial buildings. Replacing elevator counterweights with electric power generators. Install equipment that transfers the mechanical vibrations of machines to piezoelectric devices that produce electrical energy. Place piezoelectric carpets on all floors of buildings, especially where more people and machines pass. Install hydraulic turbines coupled with electric power generators in the master piping of buildings, in order to produce electrical energy with the descent of the water. Installing micro hydraulic turbines coupled with micro electric generators in the pipes of taps, showers, and toilet flush boxes, in order to transform the hydraulic force of the water output into electric energy and to inject the electric energy produced by all these devices into the public electric energy distribution network.

- **In the private domestic environment:** install propeller-less wind-powered electricity generators on top of residential buildings. Install hydraulic turbines coupled with electric power generators in the master piping of residential buildings, in order to produce electrical energy with the descent of the water. Replace elevator counterweights with electricity generators. Use piezoelectric carpets on the floors of residential buildings, especially where there is a greater flow of people, such as elevators, lobbies, corridors, etc. Installing micro hydraulic turbines coupled with micro electric generators in the pipes of taps, washing machines, showers and toilet flush tanks in order to transform the hydraulic force of the water outlet into electricity and inject the electricity produced by all these devices into the public electric power distribution network.

- **In industries:** install hydraulic turbines coupled to electric generators in the master pipe of water tanks, in order to transform the hydraulic force of the water descent into electric energy. Install hydraulic micro turbines coupled to electric micro generators in the pipes of taps, showers and toilet flush boxes, in order to transform the hydraulic force of the water output into electric energy. Installing equipment that transfers the mechanical vibrations of machines to piezoelectric devices, with the purpose of producing electrical energy. Install devices that transform sounds into electricity. Install piezoelectric carpets on floors where more people and machines pass and inject the electric energy produced by these devices into the public electric energy distribution network.

- **In vehicles:** equip all motor vehicles with piezoelectric seats and tires, and regenerative shock absorbers and brakes, to produce electrical energy, to be injected into the vehicles' batteries.

- **On streets and highways with intense vehicle movement:** install piezoelectric material and hydraulic devices between two metal plates on the vehicle lanes in order to transform the mechanical force of the mechanical impacts produced by the passage of vehicles into electrical energy and then inject the generated electrical energy into the public electric power distribution network. Hydraulic devices are used to produce an oil flow to move a hydraulic turbine coupled to an electric power generator. The highways that produce electrical power are also the suppliers of electrical power to the electric vehicles, via a overhead feeder cable installed above the road's running surface. The electric motors of the electric vehicles are powered by means of a stick (pantograph), by means of which they receive the electrical energy from the overhead feeder cable installed above the road's running surface.

- **In railroads:** install piezoelectric and hydraulic devices between two steel bars, which when pressed by the wheels of trains and subways, produce electrical energy during the passage of locomotives and wagons; and inject this produced electrical energy into the public electric power distribution network. The bottom steel bar is fixed on the railroad ties of the train and subway tracks. Hydraulic devices are used to produce an oil flow to move hydraulic turbines attached to electric power generators. The electrical energy produced by these devices is injected into the public electrical power distribution network.

These electric power generating railroads are also supplier of eletricity for the electric motors of electrified railroad and subway locomotives that are supplied with electricity by a overhead feeder cable installed above the railroad tracks. The electric motors of electrified railway locomotives and subways are powered by a stick (pantograph), through which they receive the electric power from the overhead feeder cable installed above the rails.

POSSIBILITIES TO PRODUCE ELECTRIC ENERGY

There are some possibilities to produce electrical energy to replace the energy dissipated during the process of transforming energy from one form to another form.

Sea waves and motive force

Ocean wave energy is already a developing reality, with some established projects, particularly in Australia.

Here, in this book, the concept is that it can be a mechanical system that transforms the kinetic energy of the sea waves into motive power, which in turn is transformed into electrical energy through electricity generators. The system is composed of a sub axial water wheel, whose fins are in contact with the sea waves to capture the kinetic energy of the sea waves. This kinetic energy is converted into driving force, and the driving force is converted into electricity by means of electric generators attached to the shaft ends of the sub axial water wheel. The shaft of the subaxial waterwheel is supported on bearings fixed on two floating bases, long enough to give stability to the mechanical system, The electric power generators coupled to the ends of the sub axial waterwheel shaft are fixed on the floating bases, an electric generator on each floating base. The mechanical system is anchored in the sea, and the electrical power produced is conducted by submerged cables to an electrical power transmission line on dry land. The power dimensioning of this mechanical system is made by proportionally increasing or decreasing each of the integral parts of the mechanical system. The mechanical system is made of steel covered with fiberglass, to avoid corrosion of the steel by the salinity of the sea water. It is a simple and a low cost mechanical system that can be produced quickly and in large quantities to produce clean, renewable, and sustainable electricity without polluting the environment.

Geostationary photovoltaic platforms

Many scientific researches attest that solar energy captured in space can be transformed into microwaves and sent to planet Earth. The technology has not yet been deployed, but it is feasible.

The system presented here is the installation of geostationary platforms equipped with: photovoltaic plates to produce electrical energy, a device that transforms electrical energy into microwaves and a microwave emitter and receiver device to interconnect the geostationary platforms, forming a microwave ring around the planet Earth. These microwaves are directed to the planet Earth and received by microwave receivers installed on the Earth's ground. The microwave receivers are equipped with devices that convert the microwaves into electrical energy, which is injected into the public power distribution network through an electrical power transmission line. This system allows every place on planet Earth to be supplied, without interruption, with electrical energy derived from sunlight. It is a clean, renewable, and sustainable source of energy. It is also an undertaking that transcends the nationalistic concept of each country and that, therefore, must be built and maintained with the financial contribution of all interested countries, in proportion to their energy consumption of the electric power coming from the geostationary platform and in acordance of the contribution capacity of each country.

Floating platforms with photovoltaic plates and with propeller-less wind generators

With some successful installations in Brazil, Japan, and England, these are floating platforms equipped with photovoltaic plates and with propeller-less wind generators of electricity that produce electrical energy. The floating platforms are connected to ships and generate electricity to drive the electric motors and for the ships' internal consumption. On arrival at the port entrance, the floating platform can be disconnected from the ship, towed and parked in nearby marinas, and reconnected to the ship upon leaving port.

A second alternative is to install floating platforms, equipped with photovoltaic plates and propeller-less wind generators, over the extensive surface areas of the water stored in the reservoirs of hydroelectric plants to produce electrical energy clean, renewable, and sustainable to be injected into the public power distribution grid through an electric power transmission line. The photovoltaic plates and the propeller-less wind generators can be installed on floating bases made of steel and covered with fiberglass. Fiberglass is a light, strong, and non-oxidizing material.

Electricity-generating soles

Obtaining electrical energy through some mechanical system has been a reality since as far back as 1831, with Michael Faraday and his first generator. In 2014, a 15-year-old Filipino made a homemade insole that generates electricity through the mechanical pressure generated by stepping on the ground. He succeeded in proving that "the invention could fully charge a small battery of lítio (400 amps) running for eigth hours straight.

In this way, the use of power-generating insoles in all shoes could produce electricity to be injected into a cell phone battery, or a battery to be injected into the public electric power grid. This is a potentially significant energy source, since billions of people can produce electrical energy just by walking.

USING BIOMASS AS FUEL

Sugarcane bagasse is an important source of biomass that is still little used and has great potential as a fuel to produce electricity through thermoelectric plants.

A thermoelectric plant generates thermal energy by burning fuels such as bagasse, wood, oil, natural gas, and other products. A volume of water is heated by burning some fuel, which is transformed into steam that turns a turbine that drives an electric generator. In this way, electricity is produced from the kinetic energy in the passage of steam through the turbine, transforming mechanical energy into electrical energy.1

In addition to the aforementioned sugarcane bagasse, three other types of biomass are briefly presented below: giant bamboo, water hyacinth, and algarroba tree.

A notorious source of biomass as fuel - to produce electricity through thermoelectric power plants - is the wood of giant bamboo. After three years of planting, giant bamboo reaches maturity, with thirty meters high and with thirty centimeters in diameter, and can be harvested.

The cost of the kilowatt-hour (KWh) of electricity that can be produced by thermoelectric plants that use giant bamboo wood as fuel is equivalent to the cost of the KWh produced by hydroelectric plants.

The use of giant bamboo biomass as fuel to produce electricity and activated charcoal is a clean, renewable and sustainable source of energy, therefore, it has no environmental impact, because the carbon dioxide released into the atmosphere (resulting from burning giant bamboo wood) is captured back as a result of the continued growth of new giant bamboos in the same amount of burned giant bamboos.

The use of giant bamboo wood as fuel for thermal power plants, as well as the activated charcoal generated by its controlled burning, can replace both current fossil fuels and the mineral coal used in the blast furnaces of steel mills. As a result, a lot of carbon dioxide will no longer be released into the atmosphere. The plantation of giant bamboo and the production of activated charcoal from giant bamboo wood can be located near the steel mills. The giant bamboo plantation and the bamboo-fired power plants can be located near to the electricity consuming centers. In this way, the construction of long and costly electric power transmission lines is not necessary, as is the case with hydroelectric power plants, which are usually located far away from electricity consuming centers.

The productivity of giant bamboo cultivation can be increased with the use of modern agronomic technologies and modern forest engineering technologies and thus the cost of producing electric energy with giant bamboo wood as fuel can be significantly decreased.

Another important source of biomass that can be used to produce biogas to be used as fuel for the production of electricity through thermoelectric power plants is the water hyacinth, scientific name Eichhornia crassipes.

The water hyacinth is a floating aquatic plant that feeds on organic waste mixed in fresh or brackish water. Its reproduction does not require seeds, because new shoots are born directly from the stem of another plant. However, lilac flowers with a blue background are produced, due to a biological trigger that is triggered when the water hyacinth feels threatened by a lack of nutrients or when it is attacked by herbicides. Each flower produces up to 120 seeds, which they dive into the water and wait for the right moment to germinate. The plant has underwater roots up to one meter long, full of hairs and rhizomes, and stems with leaves above the water surface up to fifty centimeters high. The water hyacinth can produce up to 1.1 kg/m^2 of new biomass daily in summer and up to 0.7 kg/m^2 in winter.

To transform the biomass of the water hyacinth plant into biogas, the following system can be built: build a reservoirs one meter deep to receive the water from the sewage system of the cities, in order to cultivate the water hyacinth plant continuously to treat the sewage of the cities, at low cost, and return the clean water to the rivers. These reservoirs do not give off a bad smell, so they can even be built inside the cities. Water hyacinth plants are harvested whole, with leaves, stems and roots, and crushed, fermented and transformed into biogas by biodigesters.

The biogas obtained from the biomass of the water hyacinth plant can be transformed into electricity through the following systems: (a) thermoelectric plants powered by biogas turbines, (b) combustion motogenerators, using biogas as fuel, and (c) directly in households, to replace liquefied petroleum gas in household stoves. These systems produce clean, renewable and sustainable electric energy, because the carbon dioxide released into the atmosphere (resulting from burning biogas or biomass) is captured back as a result of the continuous growth of new water hyacinth plants, which replace the same amount of harvested water hyacinth plants.

In addition to biogas production, the chemical composition of the water hyacinth plant of the residue from the fermentation process to be used as a raw material to produce:

1. Plant protein, for the production of protein supplement, both for people and animals.

2. Ethanol and plastic.

3. Organic compost rich in humus and mineral nutrients, which can be used in agriculture as organic fertilizer.

The Algaroba tree, with the scientific name Prosopis juliflora and also called the desert soybean, is a source of biomass to be used as fuel to produce electricity through thermoelectric plants.

The Algaroba tree is a thorny leguminous tree native to the Peruvian desert and develops in arid (desert) and semi-arid regions, with only about one hundred millimeters of rain per year and has pivoting roots to reach the water stored underground. Its trunk measures forty to eighty centimeters in diameter; the crown, eight to twelve meters in diameter; its height is six to fifteen meters; and they have roots and also pivotal roots to reach underground water. Besides flowers, it produces, from the third year of age, pods with seeds with a pleasant flavor and aroma and rich in sugars, proteins, mineral salts, and other substances, which can be transformed into raw material for the production of breads, cakes, cookies, puddings, jellies, soups, and other tasty and nutritious foods.

The pods, with or without seeds, can be crushed and fermented in a biodigester to produce biogas. Ethanol can also be extracted from the fermented liquid using a distiller, and the residue from the fermentation is a very rich organic fertilizer (which can increase agricultural productivity).

To transform the biomass of the Algaroba tree into electricity, one must: harvest the pods with seeds from the Algaroba trees; grind them, with or without seeds; and ferment them into biodigesters to produce biogas. Biogas is used as fuel in thermoelectric power plants, to drive biogas-powered turbines coupled to electric power generators. It is a clean, renewable, and sustainable source of electric energy, because the carbon dioxide released into the atmosphere (resulting from the burning of the biogas by the turbines of the thermoelectric plants powered by biogas turbines) is captured back by virtue of the continuous production of biomass by the Algaraba trees in an equivalent quantity to replace the biomass used in the biodigester.

In addition, the Algaroba tree can act in other functions, such as:

1. Reforestation of arid and semi-arid regions, such as the Brazilian Northeast.

2. Production of raw material for the manufacture of food for human consumption.

3. Production of raw material for the manufacture of food for animal feed.

4. Development of a vigorous beekeeping with the abundant flowering of the algaraba trees.

5. Production of organic fertilizer.

6. Sequestration of carbon dioxide gas from the atmosphere.

7. Release of oxygen into the atmosphere.

Today, technologies from agronomic, forestry, industrial, mechanical, and electrical engineering, as well as human and financial resources, are already available for the implementation of the circular use of energy and the production of clean, renewable, and sustainable electricity. All that is needed to add to these resources is leadership and managerial coordination.

Author: Osvaldo Dalla Coletta
Economist CRE-SP n. 6813

www.ingramcontent.com/pod-product-compliance
Lightning Source LLC
LaVergne TN
LVHW052110160826
845678LV00015B/3478